ECO DESIGN
OUTSIDE

ECO DESIGN OUTSIDE

Green Outside the House

Lorena Farràs Pérez

FIREFLY BOOKS

A FIREFLY BOOK

Published by Firefly Books Ltd. 2012

First printing

Publisher Cataloging-in-Publication Data (U.S.)

A CIP record for this title is available from the Library of Congress

Library and Archives Canada Cataloguing in Publication

A CIP record for this title is available from Library and Archives Canada

Published in the United States by
Firefly Books (U.S.) Inc.
P.O. Box 1338, Ellicott Station
Buffalo, New York 14205

Published in Canada by
Firefly Books Ltd.
66 Leek Crescent
Richmond Hill, Ontario L4B 1H1

Printed in China

This book was developed by:
LOFT Publications
Via Laietana, 32, 4º, of. 92
08003 Barcelona , Spain

Editorial Coordinator: Aitana Lleonart
Project Coordinator: Simone K. Schleifer
Art Director: Mireia Casanovas Soley
Interior layout design: Claudia Martínez Alonso
Layout: Yolanda G. Román
English translation: Cillero & de Motta

CONTENTS

SUSTAINABILITY, THE KEY TO THE GOOD LIFE

I have been following the articles by Lorena Farràs Pérez in *La Vanguardia* (one of the leading newspapers in Spain) for a while now because, for several reasons, she is a good journalist. She uses good information sources, she knows how to summarize, and she writes in a clear and understandable manner. The book that you have in your hands has emerged from these articles, which have been suitably revised and expanded.

It should be remembered that one of the greatest pleasures unique to humans is the ability to fully grasp the fundamental nature of our surroundings through ecology—through studying the relationships between living things and their environment, i.e., nature. The link between these two aspects leads us to recover that old dream of finding the good life that our predecessors dedicated so much effort to define and achieve.

Sustainability, understood in general to be a resource management system and in particular as that which affects our habitat—the buildings in which we live and work, the urban space, city and land—can be considered as the best path toward achieving the good life. The good life means living in a healthy, comfortable and pleasant environment, balancing a respect for nature with the needs of society at an affordable cost for all. The pyramid scheme that construction has become in recent years makes us forget these objectives too often.

This sustainable relationship with nature supports multiple views and possibilities. One of them is to see how our buildings interact with the four classic elements: air, fire, water and earth. Traditionally analyzed in relation to the climate of each location, these elements have determined over the centuries the forms and characteristics of vernacular architecture, to the point of distilling highly efficient building solutions through the wisdom that trial and error provides and with the satisfaction that craftsmanship brings,[1] with simple and inexpensive techniques made available to everyone.

For example, if we consider the elements of air, water, earth and fire in the Mediterranean climate, we can see that they constitute the raw material with which we achieve well-being and the comfort we seek. The basic regulatory tools of town planning and Mediterranean architecture are defined by:

- The sun—the fire element of our existence. The orientation of a building to the sun and the dimensions and characteristics of its spaces allow us to maximize the benefits of solar gain in winter and protect us from heat and humidity in the summer. With technical devices we can capture the sun and produce heat, cold and electricity.
- Wind can also be used, or cause damage: we make use of advantageous cross ventilation to cool us down in summer, benefiting from a Mediterranean sea breeze in the afternoon or creating it artificially through fans, and we protect ourselves from wind with walls.
- Protection against water is one of the characteristics that define the typical Mediterranean roof shape and angle. The slope of rooftops allows us to know the pattern of rainfall in the area, but in a country prone to drought, rainwater should be treated and reused.
- The earth supplies us with all, I repeat, all the materials with which we build our buildings and

---

1   SENNETT, Richard. *El artesano*. Barcelona: Anagrama, 2009.

public spaces. From the natural and organic to the mineral or artificial, many construction components derive from fossil fuels. There are also construction materials that challenge us to prevent or treat environmental contamination.

This book explores several materials, construction techniques and passive or active systems, all related to one of the four elements.

However, we should not forget that besides environmental values, sustainability relies on other foundations as important as the environment. Social and economic factors constitute the other pillars of sustainability. Another fourth foundation should be added to these three classic concerns: the cultural foundation. The cultural factor that drives our habits and ways of life is what allows us to link the three main sustainability factors (environmental, social and economic).

For any reader who wishes to learn more about sustainability, I advise reading a few books that are likely to guide you along the path (which will not be obstacle-free) to achieve the good life that we all aspire to. In terms of economic aspects, the writings by José Manuel Naredo[II] are enlightening. Clive Ponting[III] is highly entertaining and offers an environmental view of history. In terms of social issues, the original vision of Richard Wilkinson[IV] is a must read, and for those who want to delve into the specific contents of sustainable architecture, among many other options, is the *Un Vitrubio ecológico*.[V] I have no doubt that reading this material will enrich the reader's knowledge and lead to reflection and a call to action that will achieve a more independent and equitable way of living.

Toni Solanas

As an architect interested in fostering closer links between architecture and sustainability in the practice of his profession and through outreach, teaching, lectures and conferences, he has published such works as *La Fàbrica del Sol* (Barcelona) among numerous books on this subject. He contributed to the formation of group AuS (Arquitectura i Sostenibilitat) belonging to Col·legi d'Arquitectes de Catalunya, of which he is chairman. He also participated in the creation of the association BaM (Bioarquitectura Mediterrània). He has won the award Ciutat de Barcelona in 1987 and was a finalist of FAD awards in 1990 and DECADA in 2000. He is also a member of the team that won the first contest of democracy in Barcelona with the project Parc de l'Escorxador in 1980 (today, Parc de Joan Miró).

II Naredo, José Manuel. *Raíces económicas del deterioro ecológico y social. Más allá de los dogmas*. Madrid: Ed. Siglo XXI, 2006.

III Ponting, Clive. *Historia verde del mundo*. Barcelona: Paidós, 1992.

IV Wilkinson, Richard; Pickett, Kate. *Desigualdad. Un análisis de la (in)felicidad colectiva*. Madrid: Turner, 2009.

V *Un Vitruvio ecológico. Principios y práctica del proyecto arquitectónico sostenible*. Barcelona: Gustavo Gili, 2007. Publication prepared within the framework of the THERMIE program for the European Commission.

PRACTICAL IDEAS FOR A
GREENER GARDEN OR TERRACE

Gardens and terraces brighten up homes, but aren't necessarily
the fertile, lush life-giving havens they can appear to be. Pesticides,
herbicides and chemical fertilizers, which give plants an unparalleled
appearance, are pollutants to the environment and to people, while
irrigation, which is inefficient and uses tap water, is highly wasteful from
an ecological and financial point of view. By making a series of changes
that do not entail a great deal of work or financial investment, you can
make your gardens and terraces much greener. The first step is to plant
native species, which are better adapted to the climate and environment,
meaning they require less care and watering. After that, you have a
myriad of options to choose from, from using trees and bushes to save
on air conditioning, to making homemade compost from organic waste,
prunings and cuttings.

NATURAL SOLUTIONS AS AN ALTERNATIVE TO CHEMICAL PRODUCTS

Most products that protect plants (pesticides and herbicides) are harmful to the environment and our health. Heavy rain causes these pollutants to seep into the ground, where they filter into nearby streams and rivers, polluting the water supply. In addition, the mere production of pesticides releases thousands of tons of greenhouse gases into the atmosphere. Prevention is the best way to avoid using these chemicals, since the plants and trees cultivated with care grow to be healthy and are able to withstand problems easily. It is therefore important to bear in mind that watering during the hottest time of the day encourages fungal infections, while keeping the bases of stems well aired deters them. There are natural preparations that are easy to make, user-friendly and help plants and trees to grow. Dandelion, macerated in water and sprayed on leaves, strengthens vegetation and stimulates growth.

Well cared for and healthy vegetation is more resistant
to diseases and pests.

Vegetation plays a useful role in combating diseases and pests. Planting lavender repels aphids, while some species attract beneficial insects such as ladybugs or dragonflies, which prey on pests. In an unstable ecosystem, insect populations that were formerly balanced by their natural predators easily become pests. Ladybugs, for example, control aphids. Water is important for attracting insects and animals, such as dragonflies and amphibians, while trees attract birds. If, despite your best efforts, you do not manage to ward off diseases and pests, you can use natural solutions to get rid of them. The artichoke is one of many useful plants that can be used as an insecticide or insect repellent whether as an infusion, decoction or maceration.

To fight aphids on fruit trees, make a spray of about ½ cup (100 g) of artichoke leaves boiled in 4 cups (1 L) of water.

Water attracts insects and animals such as dragonflies and amphibians, which help to curb the pest population.

TREES IN THE GARDEN

It is now well known that trees play an important role in the environment since they process carbon dioxide and generate oxygen. Although planting just one tree does not counterbalance the carbon footprint of one home, the cumulative effects of each small effort can make a big difference. Trees and shrubs can also be a dwelling's allies: deciduous plants planted on the south side of a home provide shade in the summer, while, in winter, they let the sun's rays penetrate. On the other hand, evergreen trees and shrubs planted on the north side shield the cold wind in winter, and help keep the home cool during the summer. Additionally, planting trees and shrubs in your garden not only leads to savings in heating and air conditioning, it also increases the value of your property. A house with a garden full of trees and shrubs is always more attractive than a treeless one.

How vegetation can be an ally of the home

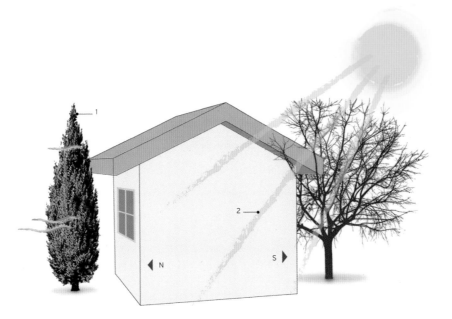

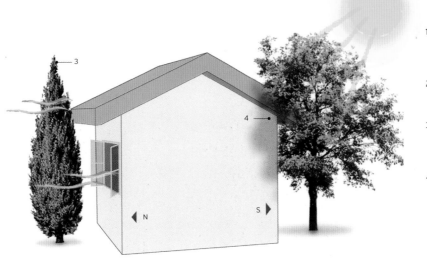

1. IN WINTER: The north face is the coldest, and vegetation, whether trees or shrubs, helps to protect the building from the wind.
2. IN WINTER: If the trees in the south and east areas are deciduous, the sun's rays can reach the dwelling, heating it.
3. IN SUMMER: The evergreen trees located to the north help cool the atmosphere by means of the process of evapotranspiration.
4. IN SUMMER: If trees are planted in the south and west, these provide welcome shade and prevent unwanted surplus heat.

ALTERNATIVES TO GRASS

A lush, green lawn requires a lot of water to maintain. While this is not a problem in areas with high rainfall; grass is not easily sustainable in drier areas. Ten square feet (3 m²) of grass consume 6.6 gallons (30 L) of water per week. Therefore, a garden of 10,000 square feet (3,000 m²) can consume as much water as a two-person household. Additionally, lawns must be mowed regularly, which in itself results in various un-ecologically sound outcomes. There are many alternatives to grass, one of the most popular of which is artificial grass; this must also be watered in summer, however, because it heats up in the sun. There is actually no need to resort to artificial solutions. Mixed lawns, for example, look similar to real grass yet require less water and maintenance. These lawns are seeded with a variety of species, such as chamomile, yarrow, clover and agrostis. On the other hand, the *Zoysia tenuifolia* grass species can be walked over because it is very durable; it needs less water and also withstands both the cold and heat. What's more, it only needs cutting once or twice a year.

Grass needs a lot of watering, particularly in areas with low rainfall, and it must be mowed regularly.

It is important to only use grass in areas where it is really going to be appreciated, such as close to a swimming pool, and to use alternatives in other areas..

WHEN AND HOW
TO WATER

Before watering, it is responsible to check whether it is actually necessary. One way to test this is by pushing a stick 12 inches into the soil near the roots. If, after removing it, the stick has soil stuck to it, you do not need to water just yet. However, if the stick comes out clean the plant needs watering. In general, we tend to overwater plants. It is preferable to water plants abundantly but infrequently to ensure the water filters down to the roots. In summer, the best time to water plants is at sunset to prevent evaporation, and in winter, in the early hours of the morning to prevent the water from freezing. The amount of water needed varies greatly, depending on the garden or terrace. Native plants, which are more adapted to local climates, tend to need less water. When designing a garden, it is therefore important to bear in mind its water requirements. It is also a good idea not to overuse impermeable paving because these prevent rainwater from being absorbed into the ground and watering plants naturally.

Cacti are typically planted in dry areas and do not consume a lot of water.

Tip for determining when to water

1. Push a stick 12 inches into the soil.

2. If the stick comes out clean, the plant needs watering.

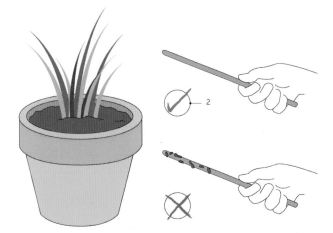

Just as important as knowing when to water, is knowing how to water. Watering by hand with a garden hose is suitable for small spaces, such as terraces, balconies and courtyards. Sprinkler or irrigation systems are used for large expanses of grass. Sprinklers are ideal for hedges, trees and bushes since they give each plant the right amount of water. Sprinklers are one of the most efficient and popular systems as they are user-friendly. However, they tend to be left on for longer than is necessary. To improve their efficiency, it is better to use ones with underground pipes and heads that sit almost flush with the ground. This way, the water passes less distance through the air and less is lost through evaporation. One way to improve the efficiency of automatic sprinklers is by installing a sensor that activates the irrigation based on the humidity of the soil, precipitation, frost and wind. Savings of up to 20% can be achieved this way.

The efficiency of the different methods of irrigation (watering can, garden hose, sprinkler or drip) varies depending on the particular usage.

USING RAINWATER

Since it contains no lime, chlorine or other chemical products, rainwater is perfect for watering your garden. Nevertheless, many homes waste their rainwater, watering their gardens with tap water, which is neither ecologically, economically nor socially responsible, especially in light of the fact that more than 10% of the world's population does not have access to clean drinking water. There are two main ways to harvest rainwater: the first, and simplest, consists of a barrel that collects the water from the roof via your gutters and downspouts. The water passes through a small, simple filter fitted in the drainpipe, which removes leaf fragments and other impurities. The barrel is typically located in the area where the water will be used, and most have a tap allowing the water to be emptied directly through a hose or into a watering can. Barrels are available in different sizes, typically from 25 to 100 gallons (95-400 L) and with different finishes, with something to suit everyone.

Rainwater storage in barrels

There are a whole range of different rainwater harvesting barrels.
These have a tap to allow the water to be directly extracted
for irrigation.

The other method available to harvest rainwater is more complex, although it is based on the same concept. The rainwater is collected from the roof by means of channels and drainpipes; it is filtered and stored in much larger collection tanks, which tend to be underground, thus preventing direct exposure to the sun. The most complex systems are even connected to a home's plumbing system, meaning that the rainwater is not only used for watering the garden, it is also used for other domestic usages not requiring drinking water, such as washing clothes and flushing toilets. The only downside to this solution is that two plumbing systems are required, one for potable water and another for the rainwater. It is also more expensive, although in areas with a high level of rainfall, savings of up to 50% in water bills can be achieved.

The water collected by means of drainpipes and channels is filtered and stored in an underground collection tank to be later used in the house or the garden.

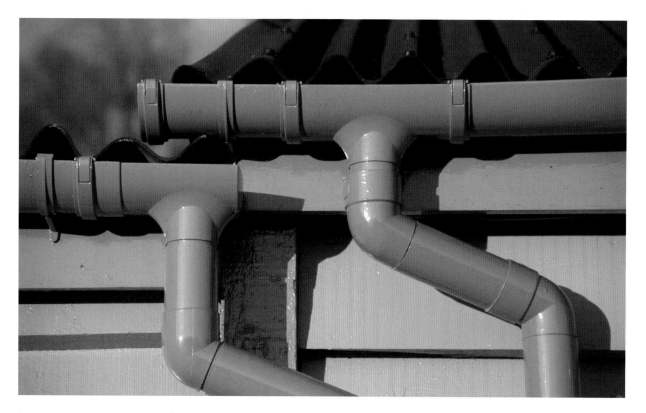

Rainwater storage in an underground tank

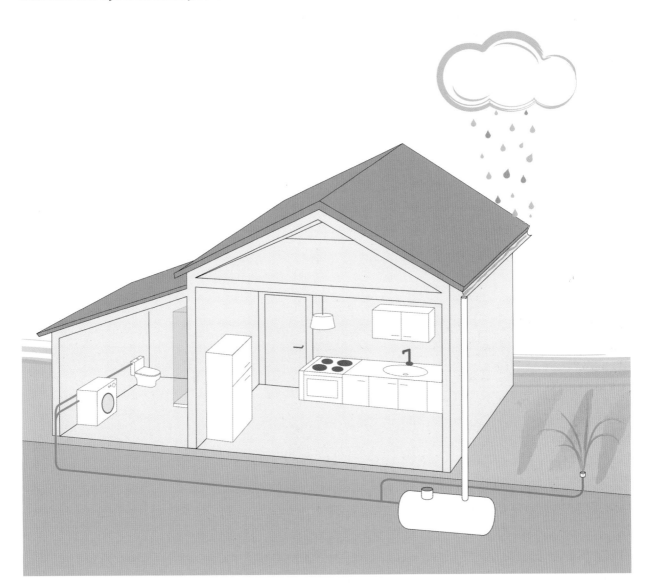

HOMEMADE COMPOST

Every day large quantities of waste are thrown away, and this could be easily reused or recycled at home. This is the case for organic or biodegradable waste, which can be used to make compost. It is mainly kitchen waste (such as leftovers from fruit and vegetables) and from the garden (such as cuttings, earth or dry leaves). However, many more things, such as paper and cardboard, teabags, coffee grounds and biodegradable diapers can be composted into organic fertilizer. Making homemade compost is not complicated. There are composters on the market that are affordable and very easy to use. In large gardens, they can be used in a shaded corner. In this case, you just have to remember to stir it around at least once a month. Homemade compost is a high quality organic fertilizer that is far better than chemical fertilizers, which can be harmful to our health, as well as the environment.

Waste that can be used to make compost

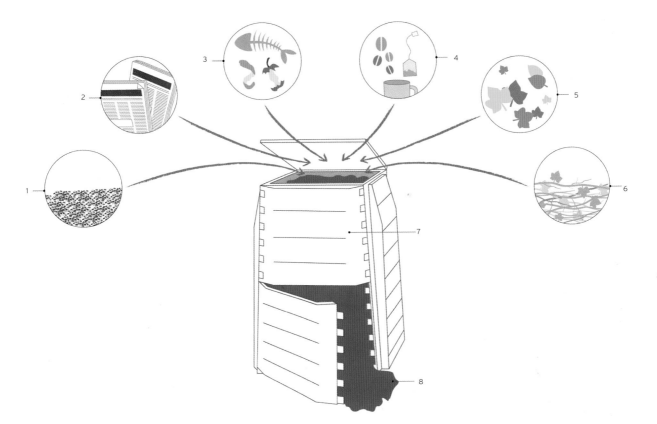

1. Ash
2. Paper and cardboard
3. Food waste
4. Tea and coffee grounds
5. Dry leaves
6. Garden waste
7. Aerobic decomposition
8. Homemade compost

URBAN ECOLOGICAL VEGETABLE GARDENS

Living in an apartment in a city is no excuse for not having a small vegetable garden on your balcony or communal yard/rooftop. Growing produce, herbs or medicinal plants is a healthy and ecological hobby that is on the rise. It is ecological because it allows us to consume produce that has been grown in accordance with environmentally friendly criteria, and it is healthy because of the properties of the products grown, which also tend to be tastier. Growing your own food at home is also a way of substantially reducing your carbon footprint. Many foods are transported hundreds of miles from their place of production before they arrive in your kitchen. With products from an urban garden, this distance is reduced to zero, with the resulting savings in energy consumption and carbon dioxide emissions. Finally, urban vegetable gardens also help to minimize waste since food grown in urban gardens is not packaged.

Features of an organic urban garden

1. Plant seeds
2. Substratum
3. Plant pot around 8 inches (20 cm)deep
4. Drainage
5. No use of pesticides, herbicides or chemical fertilizers

GREEN ROOFTOPS AND LIVING WALLS

When considering installation of a green rooftop or living wall, imagine the feeling of lounging under the shade of a leafy tree—this is the pleasant, peaceful effect the features will produce. Also, thanks to the natural processes of evapotranspiration, a process by which water is transferred from the land to the atmosphere by transpiration from plants, vegetation provides cooler environments. Soil and vegetation form the perfect insulator. Green rooftops and living walls are an excellent way of saving on air conditioning, and their advantages do not end there. Vegetation reduces pollution, absorbs carbon dioxide and promotes biodiversity. And, in the case of green tank roofs (page 36), the house gains extra living space. Both living walls and green rooftops involve extra costs. In the simplest cases, the investment can be easily recouped with the savings made in air conditioning. However, in the case of the more technologically complex living walls, the expenses are considerable, as are the benefits.

GARDEN ROOFTOP

Rooms or floors located directly beneath the roof are the hottest in summer and the coolest in winter. One solution to improve the thermal performance of rooftops is to cover them with vegetation, creating what is known as a green roof. These roofs harness the insulating effect of the earth thanks to its thermal inertia. Living beneath a garden not only reduces the costs of air conditioning and heating; green roofs are great noise insulators, and they extend the lifetime of the roof by about 40 years as they protect it from sunlight, and other eroding factors. Green roofs also reduce dust and other pollution, absorb carbon dioxide and provide a natural habitat for local wildlife. The only drawback to installing a green roof is that they cost more than a conventional rooftop, but there are alternatives to suit most budgets.

Layout of a green roof

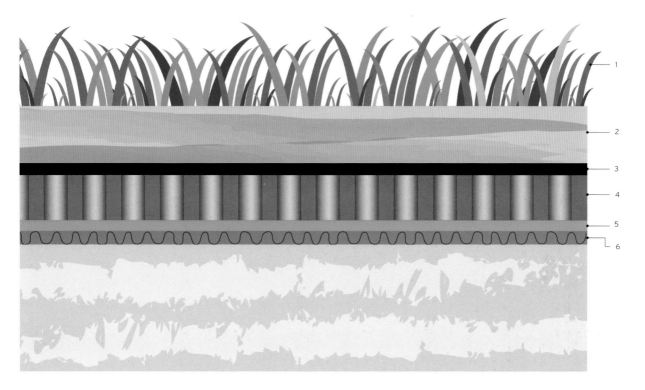

1. Vegetation
2. Earth or plant substrate
3. Geotextile (fabric that prevents unwanted mixing of soils with different properties)
4. Drainage layer
5. Anti-root barrier
6. Waterproof membrane

Generally speaking, there are two types of green roofs: extensive and intensive roofs. The former tends to be characterized by a maximum soil thickness of around 5 inches (12.5 cm) and are home to herbaceous plants, mosses or vegetation of the *Sedum* genus, succulents that are also known as "stonecrops." Intensive roofs tend to have thicker layers of substrate over which a great variety of plants, shrubs and even trees can grow if the soil substrate is a minimum of 24 inches (70 cm).

Extensive roofs do not require a great overhead for the building structure and can be installed on any roof, provided that the slope is less than 45 degrees. Intensive roofs, however, whose substrate is deeper than 24 inches, produce a great load and the building structure must be prepared to withstand this. Another important aspect is the upkeep. It is advisable to plant native species or those from the *Sedum* family, which require less watering and are highly resistant.

Example of an extensive roof growing smaller plants.

On intensive roofs, larger plants and even trees can be planted.

GREEN TANK ROOFTOP

Green tank rooftops collect and store rainwater. The method employed is quite simple: it consists of placing slabs over some supports. In the space between the slabs and the waterproof surface of the building, the rainwater is stored and is filtered along the slabs. In most cases, the water accumulated can be used to irrigate the plants on the roof itself. It can also be used to supply fire prevention systems and for any other water use that does not need to be extracted from a tap. These rooftops can be walked on because of the slabs that filter the water, allowing you to create a fantastic garden or terrace where you can take a short stroll or even place a bench to sit on and rest. As well as being another living space of the home, in warmer climates the tank on the roof gives the building thermal balance, allowing the interiors to remain cool.

As they can be walked on, green tank rooftops can also be transformed into a garden or a terrace to be enjoyed.

Components of a green tank rooftop

1. Filtration slab
2. Adjustable support
3. Water
4. Overflow
5. PVC waterproof membrane
6. Hardwearing geotextile layer
7. Regulated base support

GREEN WALLS AND WALLS THAT PURIFY WATER

Living walls or green walls are very useful in warmer regions. They tend to be used on the western and southern facades of buildings, which receive the most sunlight, thus preventing the sun's rays from hitting the dwelling directly. Moreover, the distance between the plant layer and the constructed wall serves as insulation: the vegetation cools the layer of air that is in contact with the wall of the house by means of the evaporation of water on the surface of the leaves. In this way, the surface temperature of the wall falls and less heat transfers to the dwelling. On the north facade, which is the coolest since it does not face the sun, the microclimate obtained between the two walls is ideal for ventilating the home with cool air. A classic example is ivy-covered walls. When creating a green wall, it is recommended that the plant be deciduous, so that in winter, the sun's rays reach the dwelling. It is also advisable that vegetation not grow directly on the wall, but on a parallel lattice in order to create an air chamber between the two surfaces.

Advantages of a living wall

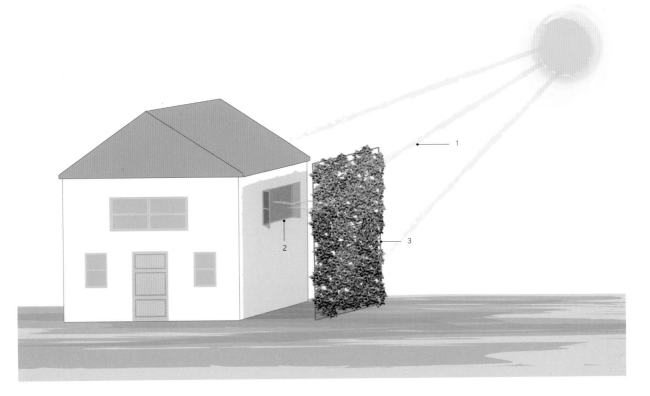

1. Protects from the direct heat of the sun's rays
2. Provides a perfect shade for ventilation
3. Insulation (reduces the transmission of heat to the dwelling)

Ivy-covered walls are the classic example of living or green walls.

More technologically complex living walls entail greater costs because they require a metal frame that binds the soil and plants. They also require more maintenance, although this depends on the vegetation planted on the wall. It is better to use native plants, but rupicolous plants (those living in rocky areas) and saxicolous plants (those growing between gaps in stones) also work well. In Europe, where ecological sensitivity has been a focus for some time, the added value of a wall capable of purifying water has been created. Domestic graywater—the water from all sources except the toilet—is pumped through the wall, which, thanks to biofiltration, is purified, collected and reused for irrigation or toilet flushing. At the same time, the graywater irrigates the vegetation on the wall.

When passing through the Babylon wall, the graywater from a home is purified to be subsequently used for irrigation or toilet flushing.

POOLS AND PONDS

Private pools are a supreme luxury that require lavish amounts of water, which is a valuable and finite resource. If, despite this, you still do not want to forego the pleasure of your own pool, it is important to try to do everything in your power to create this recreation responsibly, with a minimum of leakage and waste. There are different methods and products to minimize pool water requirements. The most important thing to remember is that you do not need to empty the pool in winter; instead, there are over-winter treatments, which conserve the water in a sanitary condition year round. To avoid using chemical treatments, natural pools are the best option. These are areas that reproduce on a small scale, the same water purification processes that take place naturally in rivers and lakes. If, rather than having a pool in which to swim, you want a water feature to enhance the landscape and promote biodiversity, a pond that purifies wastewater is your best choice.

HOW TO SAVE
WATER IN POOLS

Water is a finite resource that should not be squandered, particularly in the case of backyard swimming pools, which are a decadent luxury requiring large amounts of water. There are products and systems available to minimize energy consumption, but the savings start with the design and construction of the pool. First, you must take into account the pool's planned use in order to decide its size and make sure it is not too big. And do not forget the depth; the deeper it is, the more water it will need. The location is also important:

it must be protected from the wind in order to minimize water loss from evaporation. While in the design and construction stage, it is important to consider installing a double perimeter channel, or overflow channel, which helps to reduce usage losses from splashing. Once built, and on a regular basis, the pool should be checked for leaks. One drop per second can lead to losses of over 2,100 gallons (9500 L) per year.

Before designing and building the pool, you must decide what its main purpose will be, so as not to make it too big.

A double perimeter or overflow channel helps reduce water loss from splashing.

Filling a pool is a large investment in water; therefore, it is recommended to fill it only once. In fact, it is never necessary to drain the pool if the water is maintained in sanitary condition throughout the year. There are easy, economical over-winter treatments that prevent the proliferation of algae and bacteria that can foul your pool water. Covers are a great ally of pools all year round, not just during summer. They can reduce evaporation losses by up to 70%, prevent organic matter from entering the pool, and inhibit freezing in winter. Filters are an essential component of pools. There are low water consumption models and energy efficient pre-filtering systems. You can also install mechanisms to reuse the filtered water for other means, storing it in a tank.

Installing a cover makes for easier maintenance of your pool, and is one of the most effective measures to reduce water consumption.

It is advisable to buy low-consumption filtration equipment
and energy-efficient pre-filtering systems.

NATURAL POOLS

Although a swimming pool necessitates significant water consumption, some pools are more "green" than others. This is the case in "natural" or "naturalized" pools, which reproduce on a small scale the same water purification processes that occur in rivers and lakes, eliminating the need for chemical or artificial products. The purification process is simple. The pool is divided into two areas: a swimming area and a water garden. The water in the swimming area is pumped through the water garden, which should be equal in volume to approximately half of the total swimming area. The regeneration zone comprises a fountain or waterfall (to oxygenate the water), a gravel bed (which filters the passage of solids), and aquatic plants (the roots of which extract nutrients).

Operation of a naturalized pool

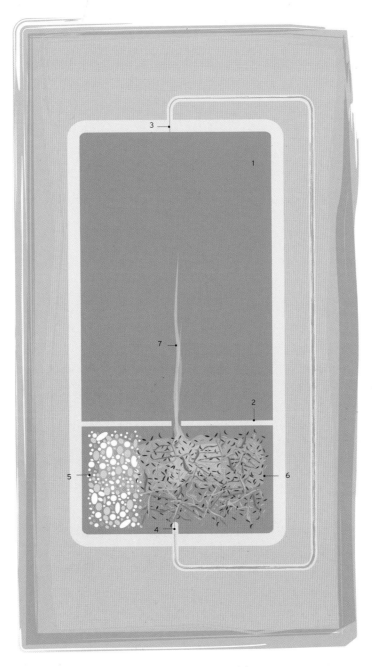

1. Swimming area
2. Water garden (should take up about $\frac{1}{3}$ of the total space)
3. The water is pumped from the swimming area to the water garden
4. Fountain
5. Gravel
6. Aquatic vegetation
7. Purified water

Swimming in a natural pool is similar to swimming in a pond or a river. Naturally, the water is not as blue as water that has been treated with chlorine, but, as it is not chemically treated, some animals can inhabit it or drink from it. These are, however, practically the only differences between this type of pool and conventional pools, because the the pool construction itself is very similar. Building a natural pool does not have to be more expensive than a conventional pool. Everything depends on the materials you use and the finish you want. However, all other things being equal, natural pools usually cost 25% more, which is the cost of building the water garden extension. On the other hand, natural pool maintenance is cheaper because you do not need to purchase chemical products. The only maintenance to consider is that the water garden requires some care, which is very similar to that of any garden.

Depending on its finished design, and especially if the water garden is placed away from the bathing area, a natural pool can have the same appearance as a conventional one.

WASTEWATER TREATMENT SYSTEMS

Natural wastewater treatment systems use and optimize a series of processes that occur naturally in the environment. They operate similarly to the process of digestion. Bars in the uptake perform like teeth, preventing the passage of thick materials. The pipes, like saliva, cause different materials to mix together. The septic tank and the deepest part of the pond are like the stomach: the compounds in the water are decomposed by anaerobic digestion.

Decomposition takes place in the upper part of the pond, an area that acts like the pancreas, where the nutrients are stabilized. The gravel and root filter acts like the liver: the roots deliver oxygen to the bacteria that live on the gravel and extract the nutrients that the bacteria produce. Finally, the vegetation lining the pond is like the intestine. Here, disinfection takes place thanks to the sun's ultraviolet rays.

The roots in the filtration area deliver oxygen to the bacteria that live on the gravel and extract the nutrients that the bacteria produce.

How a pond for treating wastewater operates

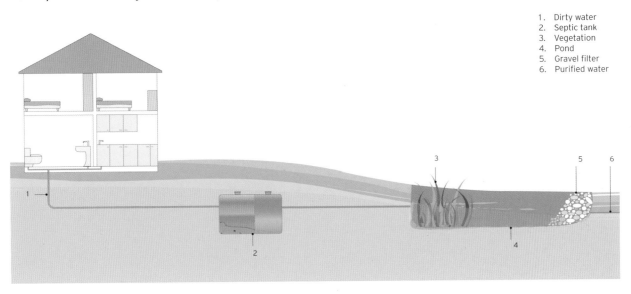

1. Dirty water
2. Septic tank
3. Vegetation
4. Pond
5. Gravel filter
6. Purified water

The different materials are mixed in the section of the piping that runs from the house to the septic tank.

These systems naturally purify wastewater without consuming any energy. The only downsides are the initial investment (the system can cost about $7,000-$10,000) and the space needed (around 65 ft²/20 m² per capita). The great advantage is that the cost is the same whether the system is designed to serve a single house or an entire residential community. In return, you obtain water that is ideal for irrigation because it contains many nutrients. The vegetation in the pond area can be used as feed or for composting. Also, when mixed with straw the sludge can be used as fertilizer. Little maintenance is required. Once or twice a year, you have to remove the pond plants and plants floating in the pond; every two to three years you must remove the bacterial sludge that accumulates in the septic tank; every ten years you must remove the sludge in the pond.

Images of a (page opposite) newly constructed wastewater treatment pond and (below) an established pond.

BIOCLIMATIC HOUSING

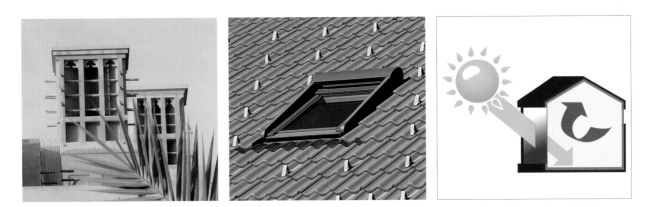

In the past, comfort was sought using natural resources. Nowadays we tend to rely solely on mechanical systems, forgetting the benefits of passive solar architecture and natural resources, such as wind, water or sun. Bioclimatic architecture seeks to build in harmony with weather rather than in opposition to it. Bioclimatic architecture is characterized by a willingness to adapt to the environment. This type of construction allows big savings in air conditioning and lighting. Just by correctly orienting the house, using insulation to its best advantage and correctly placing openings (windows and doors), you can save between 30% and 40% on energy bills. Then there are natural or passive air conditioning systems, some of which have been around for hundreds of years and offer savings without generating extra costs. The solar chimney, windcatchers, ground-coupled heat exchanger and Trombe wall are just some of these.

THE ORIENTATION
OF A HOUSE

In bioclimatic building, there are no set construction models. What works in one climate does not as a rule work in another. In a house that needs heat in winter, the main facade should be south-facing. The windows, balconies and main openings should be on this side of the house, as well as the most frequently used rooms, such as the living room or the kitchen. The goal is to harness the heat of the sun. To the east, the west and particularly the north, there should be fewer openings, and these should be smaller to prevent heat loss. In summer, the windows on the north side also provide natural cooling. In warmer climates, an opposite strategy should be adopted: the main facade and larger openings are better located on the north side, while it is better to place fewer and smaller openings on the south wall to avoid unwanted heat gain. In these regions, the heat of the sun presents a challenge.

In warm climates, the main facade is best oriented to the north and protected with shading elements.

How a house harnesses the sun in the winter and is protected from it in summer

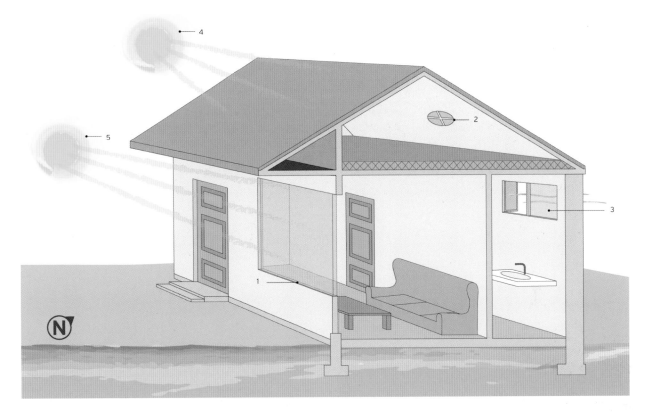

1. The main facade and the larger windows should be south-facing
2. In the east, west and north, should be a few small windows
3. The openings on the north side help cooling in summer
4. Summer sun
5. Winter sun

In colder climates, the main facade is best faced to the south to most optimally harness the sun's rays.

CROSS VENTILATION

The idea behind cross ventilation is to achieve an input and output of air to cool the home. The key to this lies in correctly orienting the dwelling and strategically locating openings that allow air to enter and exit. For this reason, cross ventilation is suitable in detached homes, although difficult to apply in more developed areas. The first step is to determine the prevailing wind directions in the area and define the layout of the openings accordingly. Generally, windows are placed in opposite facades and lobbies allow cross ventilation. On the other hand, windows letting air in should be closer to the ground (where the coolest air is located) and those letting air out should be closer to the ceiling (where the hotter air is located). It is also a good idea to have the air intake on the sides of the house with most protection from the sun, whether this be because the sun does not directly hit those sides or because they are protected by vegetation.

Before deciding on where to place the different openings that will facilitate cross ventilation, you need to know the prevailing wind directions in the area.

How cross ventilation operates

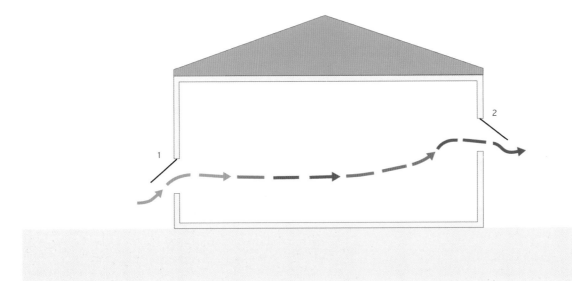

1. Cool air enters the home from openings placed low in the facade, where cool air is typically located

2. The cool air sweeps through the home exiting through openings placed higher up, pushing out warm air

SOLAR CHIMNEY

The solar chimney is a feature of traditional building. It consists of a chimney painted black that cools the house by harnessing the sun's heat. As it is painted black, a color that has a great capacity to absorb the sun's heat, the chimney heats up more quickly, as does the air inside it. This results in an upward airflow that expels the hot air from inside the house. Thanks to this passive cooling system, a cool atmosphere can be achieved on days when there is no breeze at all. The power of the movement of air should not be underestimated: it can reduce temperatures by 5 to 10° Fahrenheit (3 - 5°C). Although it is difficult for a solar chimney alone to provide a sufficient level of comfort in very hot climates, it does help to reduce costly cooling bills.

How a solar chimney operates

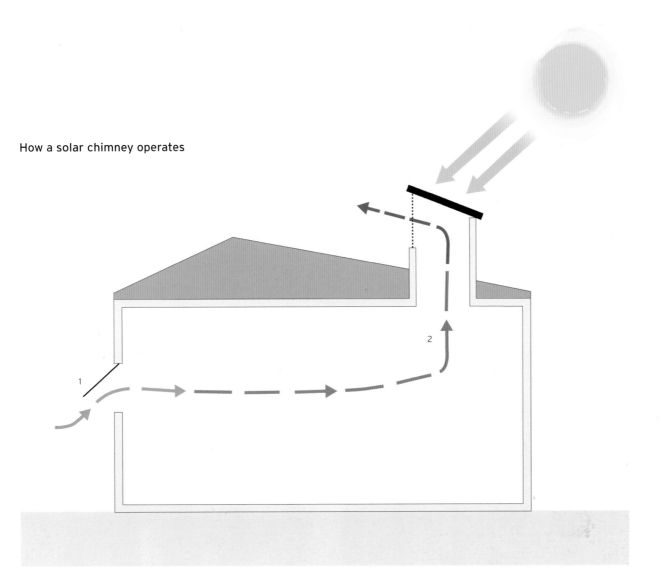

1. Cool air enters the home

2. Warm air is pulled up and out via the upward airflow created by the solar-heated chimney

WINDCATCHER

As the name implies, the purpose of a windcatcher is to harness the wind to increase the movement of air from within a house. This passive cooling system comes in the form of a small square or rectangular tower located on the roof, containing vertical bars across which the outside air blows. When passing through the tower, this hot air from the outside draws out the warm air from inside the house. This temperature management technique is mainly employed in large constructions, but it can also be used in smaller buildings. Windcatchers are often seen used in the Middle East because the feature is an Arab invention used since ancient times. Like the solar chimney, a windcatcher alone may not be enough to achieve a comfortable temperature, but it helps cool the building with zero energy costs.

A modern golden windcatcher located in Dubai. The passive cooling system helps inhabitants to withstand the high temperatures typical of the area.

How a windcatcher operates

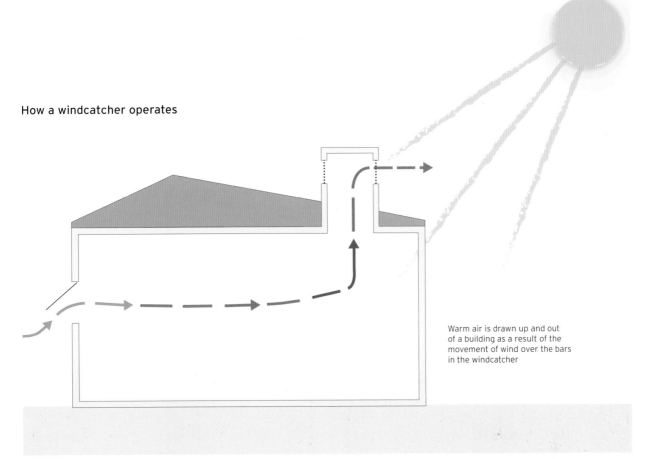

Warm air is drawn up and out of a building as a result of the movement of wind over the bars in the windcatcher

Windcatchers are commonly found on the roofs of traditional buildings in the Middle East.

GROUND-COUPLED
HEAT EXCHANGER

This system makes use of geothermal energy by heating or cooling outside air before it is vented into a home. A pipe that is open to the outside (but protected to prevent entry of water, insects and rodents) is buried in the ground at a depth of about seven feet (2 m). At this depth, the earth has a near-constant temperature, which is typically warmer than the ambient environment in winter, and cooler in the summer. In this way, the system uses heat exchange to heat and cool the home. Adding a ground-coupled heat exchanger to a new build is very economical, but not in existing houses, for which the price is considerably higher. This system has a variant in which the tube, instead of being buried, passes through a body of water.

These images featuring the construction of a ground-coupled heat exchanger show the connection of the pipe with the house, the underground circuit that the air will pass along and the outlet through which it will pass to the outside.

How a ground-coupled heat exchanger operates

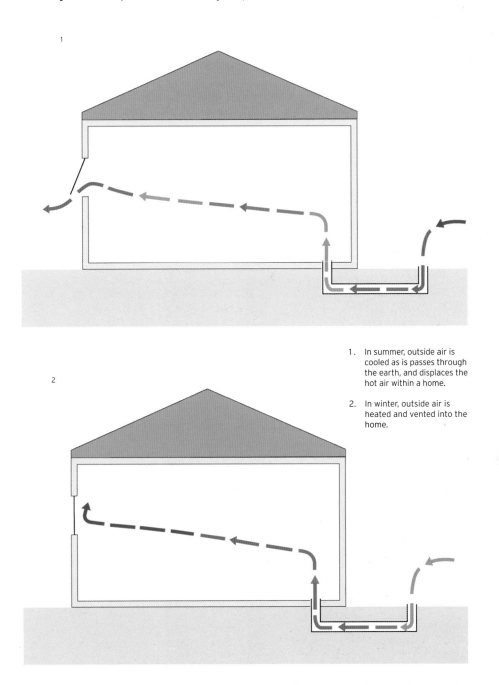

1. In summer, outside air is cooled as is passes through the earth, and displaces the hot air within a home.

2. In winter, outside air is heated and vented into the home.

INTERIOR COURTYARDS

Inner courtyards are typically found in the traditional architecture of hot regions. These courtyards provide shade for interior rooms without preventing light from entering. Their *raison d'être* is to help inhabitants to withstand high summer temperatures a little better. The efficiency of inner courtyards as a natural cooling system considerably increases if they have a water feature (such as a fountain or a small pond or a pool). The reason for this, is that as water evaporates, it draws heat from the air, producing a cooling effect. It is for this very reason that we cool ourselves down by dousing ourselves with water. Traditional buildings in the Middle East and southern Spain use this evaporative cooling system to cool homes. The cooling effect is further enhanced if there is vegetation in the courtyard.

How a courtyard with a fountain helps to cool down the home

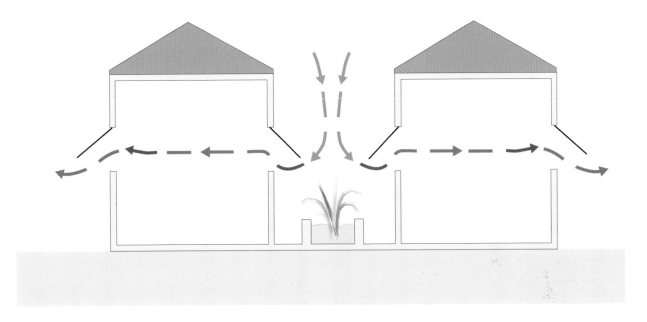

The air is drawn down and into the courtyard, and subsequently into a home, by the cooling effect of water evaporation

TROMBE WALL

A Trombe wall is a passive solar collection system that serves for both heating and cooling homes. It consists of a wall built of materials with high thermal inertia. The wall is located on the south facade, which has the highest exposure to the sun, and has a glass wall placed in front of it. The gap between the glass and the wall creates an air chamber. In winter, the sun's heat penetrates the glass and warms the air trapped within. When the upper vents of the wall are opened, this hot air enters the house. At the same time, the cool air from the interior leaves through an outlet in the base of the wall. In summer, the hot air accumulated in the chamber generates a current that draws warm air from inside the house, cooling it. Trombe walls also tend to have a visor added preventing the summer sun, which is higher, from hitting the wall directly.

The interior of a Trombe wall under construction, showing the glass wall, the air chamber and the interior wall, which in this case is white instead of black.

How a Trombe wall operates

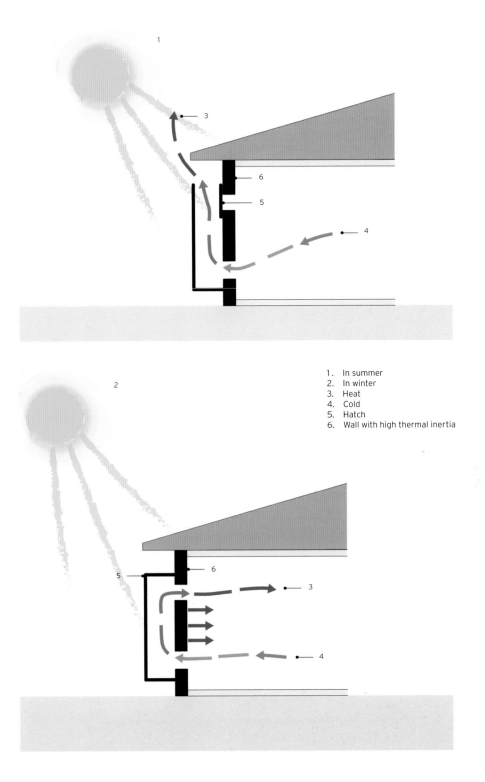

1. In summer
2. In winter
3. Heat
4. Cold
5. Hatch
6. Wall with high thermal inertia

DAYLIGHTING

Lighting represents, on average, 18% of the total electricity consumption of a home. This is quite a high amount, but it can be easily reduced. Spending less on lighting starts with designing the building with this aim in mind. Adjusting the orientation of the house and, hence, its direction in respect to the sun can generate significant savings. The first thing is to build the house with the main facade facing south, taking advantage of a greater number of hours of sunshine.

It is also a good idea to locate the most frequently used rooms, such as the living room, on this side of the house. Another factor that helps lower lighting costs is the color of walls and ceilings; lighter colors give a greater feeling of light. However, to save money, the most important factor is people's habits. Actions such as raising the blinds in the morning and not leaving lights on when the rooms are not being used can lead to savings of up to 20%.

Housing design plays an important role in energy savings. For this duplex, large windows on the south side and an open-plan layout were chosen. Therefore there are no barriers preventing sunlight from being distributed throughout the home.

Operation of a solar tube

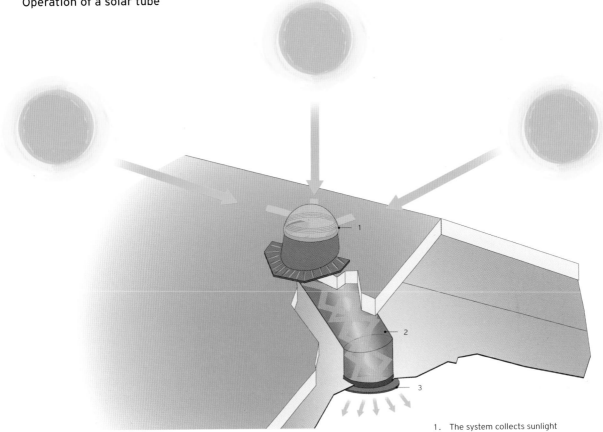

1. The system collects sunlight through a dome placed on the roof
2. The light passes inside a tube with 100% reflective walls
3. The light transported brightens up the room through a discrete circle

Light colors, especially white, reflect light. For this reason, it is advisable to paint walls and ceilings using white shades.

Most homes have hallways and corridors that are not exterior-facing and therefore do not have windows through which natural light can enter. One of the most ingenious solutions to this problem is the solar tube. A domed entry on the roof directs sunlight through a highly reflective tube into a home, lighting the space. For any rooms located directly below the roof, skylights can be fitted. These are windows in the roof or at the top of a wall. Normally, the light entering through these is more plentiful than a standard window. This means they are not particularly recommended for warmer climates, unless they include elements that prevent unwanted overheating.

SHADING ELEMENTS

The sun is the best ally of bioclimatic architecture in winter, but not in summer, especially in hotter climates. To avoid unwanted overheating caused by the sun, there's nothing like the clever creation of shade. The arrangement of shading elements (such as roof overhangs, canopies, shutters, porches, and even curtains) can prevent unwanted heat gain, reducing air conditioning costs. The openings that require the most protection are those in the west- and east-facing walls, which receive more direct sunlight in the summer. In this sense, it is important to avoid placing large windows on these sides, thus meaning there will be less surface to protect. In contrast, openings located on the south facade do not have this problem because the summer sun hits them vertically.

Of all the shading elements that exist, the canopy is one of the most efficient, cheapest and easiest to implement. Its efficiency lies in the fact that while providing shade, it also allows natural light to enter and means windows can be opened to allow air into the home. The more economical models of canopies are manually operated, but there are also canopies that can be programed to open during the daylight hours and close when the sun does not cause a problem or when it is too windy. Shutters are another great solution to protect a house from the sun. By adjusting their angle, the shutters can be placed so as to allow light to enter but deflect direct sunlight.

Canopies are a very efficient and economical solution to protect houses from the sun. Also, they are easy to fit in rural and urban dwellings.

The angle of shutters can be adjusted to allow light to enter the house, but not direct sunlight.

ALTERNATIVE IDEAS FOR DWELLINGS

Underground buildings and caves were the first homes that gave shelter to humans in prehistoric times. Architecture has evolved considerably since then, and now buildings are even available prefabricated. Prefab buildings are quick to build and highly efficient in all aspects. However, some things have not changed, such as the high quality of earth and stone as building materials because of their excellent climatic behavior and thermal inertia. In addition, cave dwellings and underground houses have an ability to blend into the surrounding environment, resulting in low visual impact. Following this philosophy, certain architectural trends are emerging that address the question of building in accordance with nature, not against it, and trying to ensure the dwellings interfere as little as possible with nature.

PREFAB HOUSING

Most of the products we consume or own have been made in a factory, and houses are no longer an exception. Sales of prefab buildings rise every year because of their many advantages and ever-improving features. Their popularity is mainly due to their low cost and speed of construction (between 6 and 8 months). However, they also have their advantages from an environmental perspective. Because much of the construction process take place in a factory, the use of materials and energy is optimized and minimized. Moreover, during this process, less waste is generated, and the low amounts of waste that are generated end up in recycling containers. Finally, there is the fact that virtually the only work carried out on-site is the connection of the different parts. This avoids stirring up a lot of dust pollution, which not only benefits the environment but also the workers and neighbors that are exposed to the construction.

Within the prefab homes sector there are different building systems. One that is currently gaining popularity is modular building. This consists of different three-dimensional modules fully finished in the factory and fitted with their systems and installations. These modules are transported to the site, where they are joined together or stacked against each other to form the final building. The final appearance of a home built with the modular building system is similar to a home built using other methods. In fact, there are modular houses that appear so similar to conventional homes that you would need to be an expert to be able to tell the difference. However, the people that do notice the difference are the owners or the tenants, who benefit from shorter lead times, lower prices, and the quality of the construction. Also, prefabrication means that a specialist workforce is employed, working in stable conditions and adhering to rigorous quality controls.

These small student apartments in Zwolle, the Netherlands, have been built using recycled shipping containers.

LIVING UNDERGROUND

Although this dwelling type is not typical, underground homes have many advantages. Being underground, their average temperature is milder and more constant than those above the earth; they are warmer in winter and cooler in summer. The ground is also a good insulating material, with high thermal inertia. That is, once the desired temperature is achieved, it is maintained for longer. Another advantage is their increased ability to withstand inclement weather or atmospheric phenomena. Their main disadvantage is problems related to moisture, light and ventilation. Proper waterproofing of walls, floors and ceilings solves the first problem. Solar tubes and a south-facing orientation solve lighting problems. For ventilation, there are several options, such as building an internal courtyard, a solar chimney or a windcatcher.

In Zurich, Switzerland, the architect Peter Vetsch built this residential complex consisting of nine houses, all covered by earth and vegetation.

This historic construction in Yan'an, a Chinese city located in the Loess Plateau, is considered to be the cradle of the country's communist revolution.

CAVE DWELLINGS

Caves have been the dwellings most used by humans throughout history. However, humans did not only live in them in prehistoric times. Caves have stood the test of time as a dwelling option, and it is easy to find modern-day examples in areas such as southern Spain, Cappadocia in Turkey, or the Loess Plateau in China. Cave houses today have the same comforts as conventional homes. Their distinguishing feature is that they are built dug into a rock, so that the rock becomes the floor, the wall and the ceiling. Their strong point is their good interior temperatures because the high thermal inertia of stone increases the thermal comfort of the home. The geographic areas most suitable for building a cave house are those with extreme temperatures, low rainfall and a soft sedimentary soil that allows for excavation, yet is compact enough to ensure the strength and impermeability of the house to prevent leakage and dampness.

The interior of a cave house in the district of Sacromonte in Granada (southern Spain), where most of the homes are built into the rock.

TREE DWELLINGS

Among the trees is undoubtedly one of the most idyllic places to live. However, this is a rather alternative option that requires more skills than one might initially think. No two trees are alike and, therefore, each project is different; although there are some general rules. First, not just any tree will do: it has to be strong enough to form a part of the house's structure. Before starting to design and build a house among the trees, the height, the shape of the branches, the strength, the species, its growth rate and so on must be analyzed. After that, the only limit is your imagination. But not anything goes. We must always strive to minimize damage to the tree and protect it from harmful long-term effects. Building a tree house does not automatically make it a more or less sustainable home. That would depend on the design, the building materials, and, particularly, its use by the inhabitants.

CAMOUFLAGED HOMES

The design of ostentatious buildings that do not take the surroundings into consideration in the least, is giving way to new trends. Buildings are becoming more modest, but also more sustainable and environmentally friendly. One of these architectural trends is camouflaged housing. The goal is to leave the smallest footprint possible and complement the landscape. So as to blend in with their surroundings as much as possible, these buildings make use of materials from the immediate environment and take advantage of slopes, terraces and forests to become an integral part of the land. Many of them have rooftop gardens planted with vegetation and have more organic forms, like those found in nature. But the most important factor here is that the building process should have the least possible impact on the environment.

ENERGY

Household energy consumption has risen continually in recent decades. We have come to own more and more domestic appliances and virtually all homes are equipped with heating systems. In addition, air conditioning systems, which consume a lot of energy, are now widespread. The result is that household energy consumption now accounts for somewhere in the region of 15% of the total energy consumption of a country, although this figure, naturally, varies by country. Households are also responsible for roughly 20% of annual carbon emissions, an amount that is expected to increase. The ideas in this chapter aim to raise awareness of renewable energy options that can be used at home and to explain how to make your home more energy efficient. By using some of these ideas, you can not only reduce your carbon footprint, but you can also save money in the long term.

SOLAR THERMAL COLLECTORS

Solar thermal energy is generally based on harnessing the sun's heat to produce domestic hot water, but it can also be used for HVAC (heating, ventilation and air conditioning), and even for heating swimming pools. The system consists of a series of plates, called collectors, which contain a liquid (usually water) that is heated by the sun. These systems are not designed to provide 100% of the hot water requirements in a home because, besides being impractical (since solar radiation is not always available), it is not economically viable because the number of collectors required would be very high. In general, they are designed to provide between 60% and 80% of hot water requirements and, therefore, require a conventional back-up system. The total cost of these installations is about $1,700-$4,000 and they are estimated to last at least 20 years.

The most widely used solar collectors are called flat-plate collectors. They can heat water to temperatures of up to 140°F (60°C), which is sufficient for domestic hot water and underfloor heating.

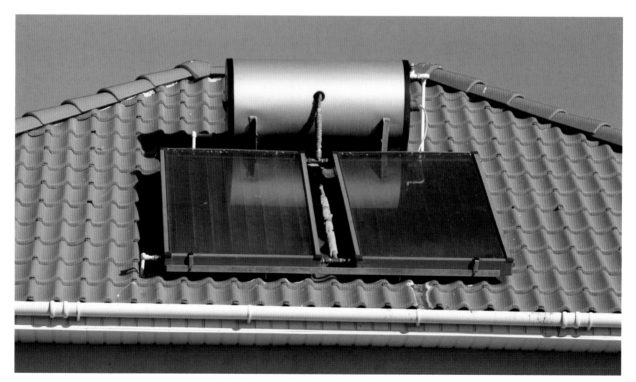

Example of installation with solar collectors

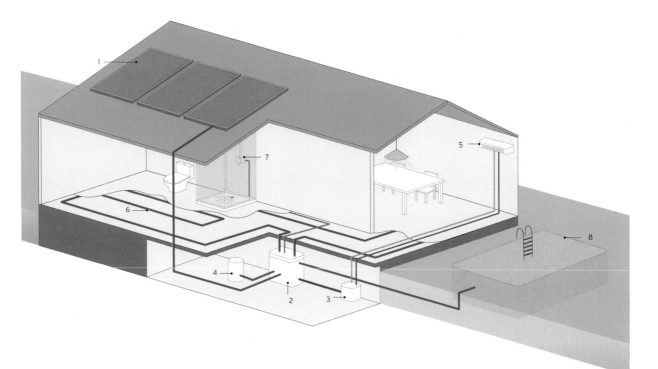

1. Solar collectors
2. Heat exchanger
3. Cooling tower
4. Auxiliary system
5. Cooling system
6. Underfloor heating
7. Supply of hot water
8. Heating for pool

Vacuum collectors (newer, yet more costly) are capable of heating water to temperatures above 140°F (60°C). They are more efficient than flat-plate collectors and are comb-shaped.

Solar thermal collectors are the most popular and efficient renewable energy system that can be found in a home. The initial investment is amortized in roughly 5 years if the fuel substituted is electricity and in 10 years if the system previously ran on gas. That is to say, after this time has passed, the energy consumed is free. Moreover, the systems have very low upkeep costs, with maintenance work mainly consisting of periodic cleaning. One of the latest developments in this field is the application of solar thermal energy to cool homes. This technology is known as solar cooling and, although available on the market, it is still very much under development. Its strong point is that the period of highest energy generation (summer) coincides with the time of highest demand. Nevertheless, the technology currently available is expensive, and adding it to an existing solar thermal system could cost upwards of $30,000.

Solar cooling does not limit you to using a particular method of cooling: it can be coupled to systems that use either air or water.

PHOTOVOLTAIC SOLAR PANELS

This technology generates electricity from the sun. Photovoltaic solar panels are particularly effective in areas with long hours of sunshine, and any surplus energy can be sold to the power grid. Installing such a system requires a sophisticated initial layout, which varies depending on the model of panels chosen and the number installed. A standard system that generates something like 1,000 kilowatt hours per year can supply 20% of the electricity needs of the average household. Very generally speaking, the cost of installing a system that meets the needs of the average home is in the area of $30,000 to $40,000. The efficiency of solar panels has improved considerably over time and today's models generate twice the electricity of those available 10 years ago. Furthermore, these systems have an estimated useful life of between 25 to 30 years, and the only maintenance required is a periodic cleaning of the photovoltaic panels.

Example of installation with photovoltaic solar panels

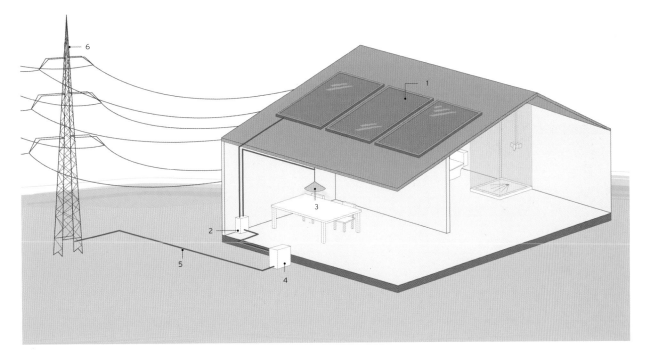

1. Photovoltaic modules
2. Converter
3. Electricity for domestic consumption
4. Counter
5. Surplus electricity
6. Connection to the power grid

For sites with no connection to the power grid, the system incorporates storage batteries that can store the energy generated during daylight hours.

MICRO WIND TURBINES

A micro wind turbine is a small scale turbine connected to a low voltage network. These very low power turbines have a maximum generation capacity of 100 kilowatts, though most are for generating less than 10 kilowatts. Wind energy for domestic use is particularly suitable for homes in remote windy areas, without access to the grid. This last point is important because turbines work best when the wind is strong and steady. Additionally, turbines should be placed away from obstacles that may obstruct the wind, such as trees. There are, broadly speaking, two models of very low power wind turbines: those with a mast, and those without (which must be mounted on roofs). Those with a mast are larger and generate more energy, and are used mainly for houses in remote areas and for domestic consumption.

In the case of dwellings in remote areas, which must generate 100% of the electricity they consume, very low power turbines are usually used in conjunction with photovoltaic solar panels.

Example of installation with wind turbine

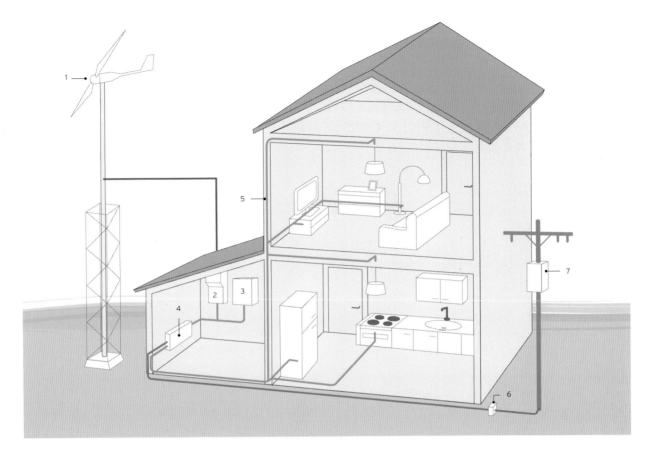

1. Wind turbine
2. Surge controller
3. Inverter
4. Battery
5. Electricity for own consumption
6. Counter
7. Connection to the power grid
    to sell the surplus energy

Roof-mounted mastless micro wind turbines are usually 6.5 feet high. Mini wind turbines with a mast may reach 65.5 feet, and large wind turbines up to 395 feet.

In urban areas, wind power is less efficient. Buildings and streets are major obstacles that create turbulence. Moreover, the wind tends to blow in gusts and from all directions. However, there are special wind turbines on the market for urban areas that counter these problems and make it also feasible to install a mini wind turbine in a city. The cost and payback period of a wind power installation varies considerably depending on the model, the capacity, the wind and consumption. On the other hand, there is also the possibility of selling surplus power to the power grid and therefore becoming a small-scale producer. The cost of electricity is therefore another factor affecting the payback period of the installation. The useful life of these installations is between 15 and 20 years and they require periodic maintenance checks.

There are special wind turbine models designed specifically for urban environments. They should be mounted on roofs, not on balconies, and it is important that the wind is not blocked by buildings or other structures.

GEOTHERMAL ENERGY

Just a few feet down, the earth has a constant year-round temperature of between 57°F and 65°F (14°C to 18°C), depending on the latitude. Geothermal energy harnesses this phenomenon to heat or cool water, which can later be used to heat or cool the home and to produce hot water. Although, to heat the water to the desired temperature, a back-up system, such as heat pumps, must be used. There are two main kinds of geothermal systems: horizontal ones and vertical ones, also known as wells. Horizontal systems require a larger surface area, but are cheaper and only need to be around 7 feet (2 m) deep. On the other hand, vertical systems, or wells, can reach up to 328 feet (100 m) deep and are more costly, but also more efficient. Another option is to run the pipes through underground or surface water, such as lakes and rivers. The installation of a geothermal energy system is 50% more expensive than a conventional system, but can lead to energy savings of up to 80%.

For new housing, the foundations are used for the horizontal circuit, while in renovated houses and urban areas, vertical circuits are more frequent.

The different types of geothermal installations

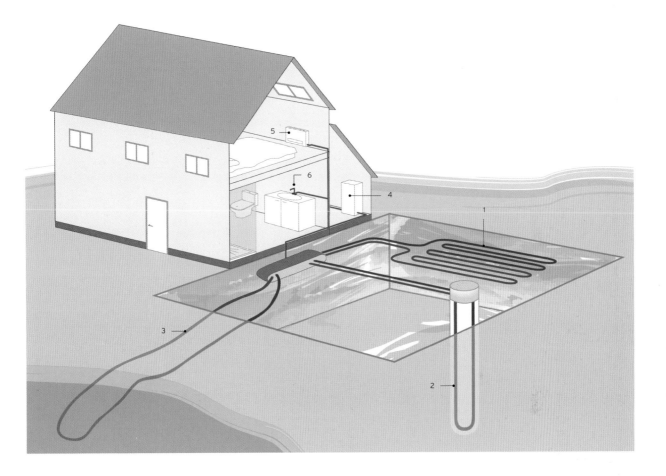

1. Horizontal circuit
2. Vertical circuit or well
3. Exchange with ground water or surface water
4. Heat pump
5. Air conditioning
6. Hot water

BIOMASS

The use of biomass to heat homes goes back to the times when man discovered fire. Today's biomass heaters mainly use pellets and chippings, and they can be used both for heating and domestic hot water. Biomass has many advantages. It is cheaper than diesel or electricity. From an environmental standpoint, the heaters release less pollution than diesel and are more efficient. Moreover, the fuel is widely available and, in the case of the pellets, gives a second life to forest and industrial timber waste. From a social perspective, using biomass is a way of caring for neglected forests and helping to manage them to reduce the risk of forest fires. The price of a biomass heating system is somewhere between $7,000 and $13,000. The only maintenance required is the removal of ash, which is not abundant and can be used for compost.

There are different models of biomass heaters, with styles to suit everyone. They work by means of electronic thermostats, and the raw materials can be introduced mechanically or manually.

The biomass cycle

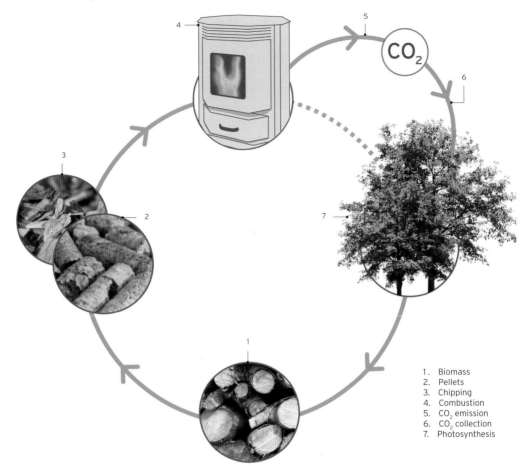

1. Biomass
2. Pellets
3. Chipping
4. Combustion
5. CO_2 emission
6. CO_2 collection
7. Photosynthesis

Pellets are small cylinders made from a pressed material similar to sawdust that are made from forest and timber industry waste. Chippings are simply a fragment of a piece of wood. Unlike pellets, they do not undergo any transformation process, but they have a lower caloric efficiency.

INSULATION

One of the best investments you can make at home to save energy on heating, ventilation and air conditioning (HVAC) is to properly insulate walls and ceilings. Insulation acts as a barrier against hot or cold air from the outside and prevents possible leakages from inside. In fact, insulation is a building's first line of defense against the climate. The amount of energy needed to maintain a home at a comfortable temperature depends largely on its level of insulation. A poorly insulated home consumes more energy because, in winter, it cools quickly and heat leakages occur, while in summer it reaches a high temperature and heats up quicker. By insulating your house well, you can save energy and cut costs by up to 30%, both in heating and air conditioning. It is estimated that the initial investment can be recouped in 5 years or less.

Most new houses come with a built-in layer of insulation, but the same is not true for many older buildings. Some states grant subsidies for the energy-related revovations of buildings.

Heat loss

A poorly insulated house loses heat in winter, so it takes more energy to keep it at a comfortable temperature.

The outer shell of a building is where it loses or gains heat if not well insulated. For this reason, attics or rooms directly under the roof tend to be the coldest in winter and the warmest in summer.

PROPERLY SEALED WINDOWS

It is estimated that the windows are responsible for generating between 25% and 30% of the HVAC needs of a home. If poorly insulated, in summer they are a source of unwanted heat and in winter, while hot air escapes from the inside, cold enters from the outside. So investing in window insulation is a good way to increase your energy efficiency and to save money. The thermal insulation of a window depends on the quality of glass and type of frame. One of the most popular solutions is double glazing, which acts as a thermal insulator while also blocking out noise from the outside and reducing condensation. Also, the best frames are those built using a system known as thermal bridging, which contain insulating material between the inner and outer frame. It is further advisable to cover cracks and reduce drafts in windows and doors by using silicone, putty or draft excluders.

Wood is an ideal material for frames because it is a great thermal and acoustic insulator, as well as being natural, biodegradable and renewable. Conversely, iron and aluminum let in the cold and heat.

Double glazing traps air between two panes of glass,
creating an insulating barrier that reduces heat loss.

In addition to almost halving the heat loss of single glazing,
double glazing reduces drafts, condensation and frost.

OUTDOOR LIGHTING

Reducing the costs of outdoor lighting and the amount of light pollution, which affects wildlife, requires minimal investment. The first thing to do to save energy and money is to opt for energy saving light bulbs. Incandescent light bulbs only convert 5% of the total energy consumed into light, while the remaining 95% is lost as heat. On the other hand, energy-saving light bulbs not only consume 80% less, but also last at least six times longer. Another way to save is to put a visor on lamps or to buy models with a built-in visor. This prevents the light rays from being projected upwards into the sky and ensures that the light shines where it is needed. The result is that it takes less energy to light an area and helps to reduce light pollution. Lights can also be fitted with timers or movement sensors. Solar lights are another option: these charge during the day by means of a solar panel and light up in the dark.

The visor on the lamps prevents the sky from being lit up and directs the light where it is needed. In this way, you can light an area consuming less energy and reducing light pollution, which has an effect on wildlife.

Energy-saving light bulbs are more expensive than incandescent ones, but the higher outlay is quickly recouped owing to their higher efficiency and longer lifetime.

Solar lights do not need a connection to electrical outlets: they charge during the day by means of a solar panel.

GREEN MATERIALS

Most materials used in the construction of homes are not very environmentally friendly or good for our health. They are manufactured using industrial processes that release large volumes of carbon dioxide and require huge amounts of energy. Additionally, after their useful life, they are not biodegradable, and become waste that is difficult to integrate into the environment. This is why it is important to choose the materials that will form part of a home carefully, from the very first stone to the paint. The most ecological or sustainable option is to try to choose materials that are as natural as possible, such as mud, straw or wood. It is also important to choose materials that are locally available, and avoid the negative environmental impact caused by the transportation of goods, while also helping to sustain the local economy.

MUD

It is estimated that between one half and two thirds of houses worldwide are made from mud. The appeal of clay is that it is abundant, locally available, and cheap, aside from being a first-class construction material. Besides being 100% natural and requiring virtually no energy for its transformation, clay is a good acoustic and thermal insulator and allows moisture to pass through, letting the house breathe. And, being natural, after its useful life, it does not become waste. There are two main mud construction techniques: tabby (a mixture of earth, sand and pulverized sand) and adobe. The first involves constructing a formwork that is compacted at the same time as it is filled so that the mud is uniformly and firmly packed. Adobe consists of mud bricks dried in the sun. As these bricks are not fired, unlike cement bricks, the earth does not lose its thermal inertia.

The finish of a house made from mud is very similar to that of a conventional house and, if the structure is well made, it can last 100 years or more.

Adobe blocks (which are made from a mixture of clay, sand and water) are dried in the sun for 25 or 30 days, so the energy required to make them is minimal.

The tabby technique consists of constructing a formwork into which clay is placed, then flattened and compacted.

The only critical point of mud is water, but this is solved by laying good foundations and having an overhanging roof to protect the walls from water.

STRAW

This agricultural byproduct has excellent qualities as a building material. First of all, straw is 100% natural, so at the end of its useful life it does not become waste because it is biodegradable. It is also a locally sourced product, particularly in rural areas, and requires no prior processing. It is also an abundant and cheap material. As a construction material, straw bales are great thermal and acoustic insulators, and are also breathable, so the air inside is constantly renewed. The construction process is relatively simple. Building the walls is the quickest and easiest part. Broadly speaking, the technique consists of stacking bales of straw on top of each other, as if they were giant bricks. The trickiest part is finishing the inside layer, which is slower and must be installed with great care to protect the straw.

Straw's number one enemy is water. It is important to prevent the bales from becoming damp and to cover them properly to ward off problems in the future. With proper maintenance, a straw house can last for more than 100 years.

An example of a house made from straw under construction and completed (below). Because of the size of the bales, the walls are roughly 20 inches (50 cm) thick.

WOOD

Although it may seem contradictory, because using wood involves cutting down trees, timber is one of the most environmentally friendly materials. Wood is not only warm and inviting, but also natural, safe, renewable, reusable and biodegradable. As a construction material, it is lightweight, easy to work with and very strong. It is also a great thermal and acoustic insulator, is breathable and regulates moisture effectively. To be consumed sustainably, there are certain rules to be followed. First, opt for wood available in the vicinity in order to minimize the energy consumption and carbon dioxide emissions produced from transporting it. It is also better to avoid using fine hardwoods because these trees take a long time to grow and older forests store more carbon than the younger ones. And, finally, make sure that the wood is a product certified by the Forest Stewardship Council (FSC), which ensures that wood is sourced from sustainable forests.

In choosing wood, you contribute to the health and management of forests, which are major consumers of carbon dioxide.

The high versatility of wood, which can be used for house frames, walls, floors and furniture, is one of its major assets.

BAMBOO

This is one of the most used building materials for promoting sustainable architecture. Its top advantages are the rapid growth of the plant, which naturally regenerates every 6 years, and its flexibility and great strength. But, bear in mind that if it has to be imported, it is no longer an ecologically responsible choice of material because of the energy costs and carbon dioxide emissions that would result from its transportation. It is also important that the bamboo is certified by the Forest Stewardship Council (FSC). Besides being natural and biodegradable, bamboo is an economical material and is used frequently for building in areas where it grows. It is used as a raw material for flooring, coverings, panels, furniture, and so on. However, it has two major threats: fire and biological attacks. Therefore, it is essential to subject the canes to a special treatment in order to protect them. Its useful life is roughly 20 years.

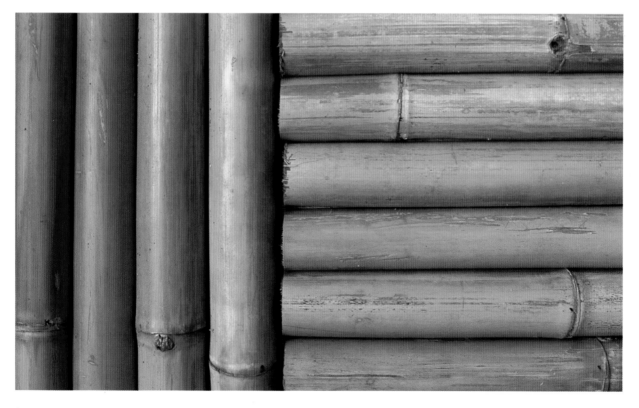

China is one of the countries where the use of bamboo as a building material is a tradition going back a long time. In these photos, this building in Beijing has been partially constructed using bamboo.

STONE

When we think of stone, the first thing that comes to mind is its hardness and durability. These are two of the most significant advantages of this building material, but not the only ones. Another of its strengths is its excellent properties as an exterior, acoustic and thermal insulator. Stone houses do not leak heat in winter, while in summer the stone absorbs solar radiation and prevents the heat from being transferred to the interior. The versatility of this material, which serves equally well for foundations, walls and interiors, is another of its great advantages. Of course, not all types of stone can be used for everything. But it also has some drawbacks. Stone is only recommended if it is available in the vicinity, and we must take into account that quarrying has a significant environmental impact. Moreover, building with stone is slow and more expensive than with other materials.

Removing stone from quarries has a major impact on the environment, as does its transport.

Stone is a traditional building material in mountain areas because of its excellent thermal inertia.

The versatility of stone means it serves equally for entrances as well as for foundations, walls and interiors.

NATURAL INSULATORS

The main natural insulation materials are cork, plant fiber (such as hemp or flax) and wood fiber. Cork is available in the form of shavings to fill cavities and in boards, but can also be sprayed. Thermal blankets can be made using the fibers of hemp or flax, two fast-growing plants. And finally, panels can also be obtained from wood fiber. Although it is true that to be transformed into insulators, these materials must undergo an industrial process, the energy required for this is much lower compared to that required for non-natural insulators. Additionally, these materials are renewable because you can always plant more trees or plants, which, in turn, help to absorb carbon dioxide. And, as they are biodegradable, at the end of their useful life they do not become waste. Natural insulators end up being somewhere between 30% and 40% more expensive, but all costs are recouped in just a few years owing to HVAC savings.

Cork insulation panel.

Wood fiber panel.

Thermal blanket made from plant fiber.

RECYCLED INSULATORS

Spending on insulation is always a good investment for both the environment and your pocket, but even more so if you choose insulation made from recycled materials. Wool scraps from the textile industry, used paper (mainly from newspapers) and waste from the timber industry are the main types of waste used to make recycled insulators. Wool, cellulose fiber and wood shavings are every bit as good as conventional insulators. They all have high acoustic and thermal insulating values. They are also excellent moisture regulators, as well as being breathable, in the case of wool and wood shavings. From an environmental standpoint, their value lies in the fact that they give waste a second life and are biodegradable. And the energy required to manufacture them is less than that required for other insulators, with the only snag being that they cost somewhere between 30% and 40% more.

Used paper, mainly from newspapers, is shredded and turned into cellulose fiber, which is used to make thermal blankets.

Scraps of wool from the textile industry and surplus wool from livestock farms are used to make thermal blankets.

Wood shavings discarded by the timber industry are reused to make insulation panels.

PAINT

The leading paint brands have lines of green products, which have been fabricated using environmentally friendly processes. But in the world of paint, there is an option even better for the environment and our health: natural paint. These paints are made from products with a mineral or vegetable origin. Their major plus point is that, being of natural origin, they are not harmful to the environment or to humans. There are natural paints for all kinds of surfaces, and these are not restricted to just standard paints: there are also enamels and rust-proofers, among others. Compared to conventional paints, these products are more breathable, they are fireproof, require fewer coats, do not stain as easily since they repel dust, and they do not contain toxic components. The only drawbacks are that there are fewer colors to choose from and they are more expensive to purchase, but in the long run, they are more cost-effective.

There are natural paints on the market for all kinds of surfaces (both indoor and outdoors, wood, metal, etc.), and not just standard paint: there are also enamels and rust-proofers, among other products.

There are fewer natural paint colors available than conventional ones, but this is not really an issue because you can always blend colors to get the one you want.

Natural paints are ideal for facades because fewer coats are needed, they are more breathable and they repel dust and air pollution, which means they stay cleaner.

Mineral-based natural paints (known as silicates) are mainly used for painting walls. Their main feature is their durability, especially on exteriors, because of their dust and pollutant repelling properties. It is estimated that one single coat can last more than 100 years. Vegetable-based paints are derived primarily from plant oils and resins, and from waxes and mineral pigments. These are used on wood. Their great advantage is their greater flexibility over conventional paints, which, over time, facilitates the upkeep of the wood and keeps it in better condition. As for their durability, they are similar to that of conventional paints. Another important factor to consider is the color to paint exteriors. In warm climates, it is advisable to use white to avoid heat gain. On the other hand, in cold areas it is better to go for dark colors, which absorb heat.

Vegetable-based natural paints are used to paint wood. Their flexability is their great advantage over conventional paint, which is less flexible and eventually cracks and degrades.

On exteriors, one single coat with mineral-based paint can last over a century because the finish is highly resistant to weathering. Interiors, however, must be painted as often as with conventional paint because the walls are dirtied by usage.

White walls are typical in hot climates, such as in southern Spain or the Greek islands. Light colors, particularly white, help prevent heat gain from the sun.

TECHNOLOGY

New technologies are increasingly present in modern living, and they are here to stay. The most advanced version of new technologies in the home is the digital home or home automation. This not only facilitates many everyday tasks, but it also makes a home more efficient, reducing energy and water consumption. However, not all technologies are beneficial. Lawnmowers or outdoor gas heaters, consume a lot of energy and release carbon dioxide into the atmosphere. Finally, this technology conceals invisible dangers. Some electronic devices, like other appliances in the home, are sources of electromagnetic pollution that can affect our health. Homes, as well as being liveable and sustainable, must also be biocompatible.

THE DIGITAL HOME OR
HOME AUTOMATION

The concept of home automation refers to the automation and control of different installations and equipment in a home, such as the lighting, air conditioning, blinds, awnings, irrigation systems and so on. For example, an automated home may have a program, called Good Morning, that is activated at the push of a button to automatically open all the blinds, awnings and windows in the home. These elements and installations can be controlled by various devices. The most common way is via portable or fixed terminals or via a remote control. They can also be controlled remotely from outside the house by using a mobile device with an internet connection. Remote control access allows, among other things, friends and family to gain entry without someone having to let them in.

Blinds, windows and awnings are just some of the elements in a home that can be automated.

Weather stations provide the digital home with the information it needs to know, for example, if the awnings should be closed because it is too windy, if it is necessary to lower the blinds because a storm is on its way, or if it is time to irrigate.

There is no doubt that the digital home or home automation increases our quality of life, but this is not the only advantage. It also allows us to save water and energy. Automating certain actions is very helpful for generating savings provided that we know how to get the most out of the technology. You can also install solar radiation sensors and program the operation of blinds and awnings so that in summer they lower automatically when the sun hits them, for example. This can save you up to 30% in air conditioning costs. Home automation systems also have an assortment of features designed to monitor water and energy consumption. This not only allows us to know the actual consumption, but also to detect system malfunctions and possible leakages. Home automation is by no means an elitist product. Fitting a basic system increases, on average, the final construction cost of a home by 1%.

Devices that monitor water consumption and energy (like electricity meters) provide information on the actual consumption of the house, which allows you to take actions to reduce it.

Different control devices are available. The most common are fixed terminals, which are often located in the living room.

The most successful digital devices are those related to security, whether these are fire or intruder prevention systems.

THE LAWNMOWER

The choice of lawnmower depends largely on the characteristics of a lawn, particularly its size. However, we must take into account that, from an ecological perspective, the best option is the manual lawnmower, which does not require any power source other than human force. If you choose a lawnmower with a motor, this should preferably be electrical, since these models are much less polluting than their gas counterparts. And, of all the models available, the least environmentally friendly ones are riding lawnmowers. These devices are highly polluting, more so than any car traveling at full speed. Another alternative is the robotic lawnmower powered by solar energy. These devices are fully automated; you just have to program them. Most are hybrid systems and, in addition to having solar panels on the top, can also be plugged into the electrical outlets.

These models are very comfortable, but also more expensive and polluting. From an environmental point of view, they are the worst option, and they are only recommended for large expanses of lawn.

Electric lawnmowers are more efficient than gas ones and less polluting. Their main drawback is the need for a cable to power the motor, which hinders mobility.

Although they are more powerful than electric lawnmowers, gas lawnmowers are also more polluting.

Manual lawnmowers require a greater physical effort from the person operating them, but they do not consume energy, so they are the greenest option.

There are robotic lawnmowers on the market powered by solar energy, such as this one by the brand Husqvarna.

OUTDOOR HEATERS

Patio heaters are energy guzzlers. As is their function, they directly expel heat outdoors and, therefore, are very energy inefficient. Most run on gas, a non-renewable resource. But the worst part is that in general, they are not equipped with filters to reduce the greenhouse gases produced by burning fuel. One possible alternative to gas heaters are electric ones that release radiant heat. Unlike gas heaters, these devices do not burn any fuel, so they do not release carbon dioxide into the atmosphere. Electricity is always safer than gas, but it is also more costly. The most environmentally friendly option of all is to have no heater and to cover up a little more if you want to enjoy the garden or terrace in winter. A good blanket can be a great help to keep you warm.

The least polluting option is electric heaters, which release radiant heat and do not burn fuel.

Gas heaters are the most widespread option, but also the most polluting.

ELECTROMAGNETIC POLLUTION

The home is a refuge for most people. However, homes often conceal unseen threats that are harmful to our health. Materials used in construction and design greatly affect the degree of health or biocompatibility of a building. For this reason, it is recommended that materials and paints be as natural as possible. Hidden dangers can be found outdoors as well. Electromagnetic pollution caused by power lines and mobile phone or telecommunications towers is the most notorious. However, nature also has its own threats. Experts in the field advise that geobiological studies be conducted prior to constructing a building to analyze possible sources of pollution emanating from the ground. These studies are particularly recommended in areas where there are faults and underground streams, because of the toxicity of radon gas.

Requirements of a biocompatible house

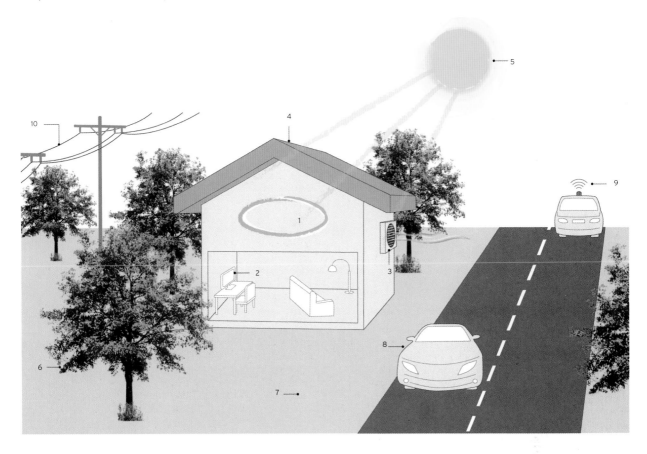

1. **Quality of the inside air**
   Keep a house well-ventilated to avoid the accumulation of chemical substances that may be harmful to the occupants' health.
2. **Electrical systems**
   Avoid radiation from Wi-Fi networks and from certain electronic equipment.
3. **Equipment maintenance**
   Without proper maintenance, some equipment, such as air conditioning units, can be a hotspot for the cultivation of bacteria, fungi and other microbes.
4. **Biocompatible materials**
   Materials must promote the health of inhabitants, be breathable, have low radioactive emissions and no toxic chemical compounds.
5. **Daylight**
   Sunlight is important for its anti-bacterial and purifying properties and to prevent vitamin D deficiency.

6. **Positive effects of plantlife**
   Outdoor air quality is better if there are wooded areas, or parks with abundant vegetation nearby.
7. **Geological radiation**
   Before building, it is a good idea to study the energies emanating from the ground and avoid areas with high terrestrial radiation resulting from the toxicity of radon gas.
8. **Site of the house**
   When choosing the site, bear in mind the nearby sources of environmental pollution, such as industry or busy streets.
9. **Noise pollution**
   Noise can cause serious negative psychological effects and other physiological problems, and, in the most serious cases, may damage our hearing.
10. **Electromagnetic pollution**
    It is important to keep a distance from sources of electromagnetic pollution such as power lines or telecommunication towers.

Human activity is another major source of pollution that directly affects the quality of the air surrounding a building. In this sense, it is best to avoid areas with busy streets or near roads or industrial areas. If no other site is available, you can design and build the house taking these sources of pollution into account. To do this, you must bear in mind that bedrooms need the most protection, as this is where you spend the most time. An ideal location is one with abundant plantlife in the vicinity. Vegetation regulates air quality, and there are even studies that show that people living in areas with plantlife live longer. And we must not forget the interiors of homes.

Homes contain many different threats, but most are an essential part of today's lifestyle. Experts recommend that, at the very least, we should avoid placing radio alarm clocks, phones or sockets by our bedside.

Some studies have found that people living in areas with abundant vegetation live some five and a half years longer than those living in areas with little or no vegetation.

The noise pollution surrounding our homes may also adversely affect our health.

The site of a house is one of the most important factors to consider before building or buying. If possible, you should avoid areas near sources of pollution.

THE THREE Rs OF HOUSE CONSTRUCTION

The current lifestyle predominant in the West produces ever-growing mountains of waste. This is also the case in the field of housing construction and demolition. In renovations, and especially in the demolition of buildings, a huge amount of waste is generated. To combat this, we need to use the famous Three Rs: reduce, reuse and recycle. The first and most important of the three Rs is to reduce the amount of waste. Reducing means using materials that, after their useful life, will not become waste, something that is not always possible. This is where the other two Rs come into play. Although not everything is recyclable or reusable, many building materials can have a second life, such as earth, rubble, iron, concrete or wood. For this, it is essential to properly separate the waste on-site.

REDUCE

The first and foremost of the three Rs is to reduce, i.e., to try to minimize the amount of waste. In the case of the construction sector, reducing means using natural materials. At the end of its useful lifetime, a natural material that has barely undergone transformation and is still in its natural state, does not become waste. Materials such as stone, wood that has been treated with non-toxic paint or varnish, straw or mud fall into this category. These are biodegradable materials and, therefore, may be returned to nature without any pre-treatment. Another way to reduce the amount of waste is to choose longer-lasting products and materials. The longer it lasts, the longer it will take to become waste and to have to be replaced. In this sense, reducing is completely antagonistic to the "disposable" model.

At the end of its useful life, wood does not become waste because it is a biodegradable material.

Some non-biodegradable materials, such as concrete, can take more than 100 years to decompose.

Thanks to its durability and strength, stone is a building material with a long useful life.

REUSE

The second most important R is to reuse. This involves trying to get the most out of a material or product before it becomes waste. In the construction sector, this means completely retrieving those components that can have a new use or even the same one, but only with a minimal transformation process. The frames of doors and windows, overlay flooring or floating flooring, tiles or even prefabricated concrete elements fall into the category of reusable materials. Most of these elements can be used in other constructions after carrying out some minor adjustments to them, provided that the base material is good quality and in good condition. There is also the more artistic option, which consists of making furniture or decorative objects with what for many is waste. For example, with a little skill and patience, you can make a shelf with leftover drywall.

With a little knowledge of carpentry, you can make furniture from old wood, such as these simple chairs.

Materials in good condition and even some complete building components, such as window frames or doors, can be reused.

RECYCLE

When generating waste is unavoidable, there is still the third R: recycle. Most construction waste is recyclable. However, it is essential to properly separate the waste on-site. Waste can be separated into three groups: inert, hazardous and non-hazardous. From among the inert materials, the most abundant is earth. If this is clean, it can be taken directly to its new use and, if not, it simply needs cleaning. Ceramic and concrete remains are crushed to obtain aggregate, which is used in road building or to make concrete. Iron can be melted and wood shredded to become chipping for biomass or a raw material for chipboard. Non-hazardous waste that is not recyclable ends up in controlled landfills. And finally, hazardous waste, such as sheets of asbestos and cement sheets, are treated and stored in special containers.

From crushed concrete, brick or ceramic waste, a granular material is obtained, which is used in road building or to make concrete.

Management of construction waste

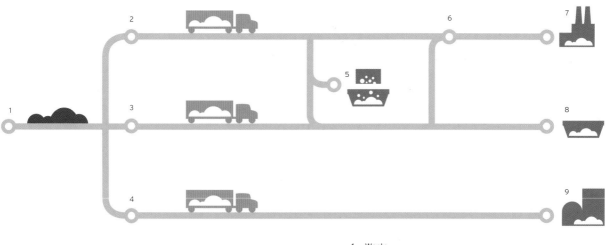

1. Waste
2. Inert
3. Non-hazardous
4. Hazardous
5. Recovery and recycling in the same center with a mobile plant
6. Plant (authorized manager)
7. Recovery and recycling in a waste-to-energy plant
8. Controlled landfill site
9. Processing and storage

Asbestos, now banned in many countries because of its high toxicity, is a hazardous waste and should be removed by specialist companies.

IMAGE CREDITS

INDEX